Menu Marketing and Management
Exam Prep Guide

Upper Saddle River, New Jersey
Columbus, Ohio

DISCLAIMER:

The information presented in this publication is provided for informational purposes only and is not intended to provide legal advice or establish standards of reasonable behavior. Customers who develop food safety-related or operational policies and procedures are urged to obtain the advice and guidance of legal counsel. Although **National Restaurant Association Solutions, LLC (NRA Solutions)** endeavors to include accurate and current information compiled from sources believed to be reliable, **NRA Solutions**, and its **licensor, the National Restaurant Association Educational Foundation (NRAEF)**, distributors, and agents make no representations or warranties as to the accuracy, currency, or completeness of the information. No responsibility is assumed or implied by the NRAEF, NRA Solutions, distributors, or agents for any damage or loss resulting from inaccuracies or omissions or any actions taken or not taken based on the content of this publication.

Sample questions are designed to familiarize the student with format, length and style of the examination questions, and represent only a sampling of topic coverage. The performance level on sample questions does not guarantee passing of a ManageFirst Program examination. Further, the distribution of sample exam questions with their focus on particular areas of subject matter within a ManageFirst Competency Guide is not necessarily reflective of how the questions will be distributed across subject matter on the actual correlating ManageFirst exam.

Visit www.restaurant.org for information on other National Restaurant Association Solutions products and programs.

ManageFirst Program™, ServSafe®, and ServSafe Alcohol® are registered trademarks or trademarks of the National Restaurant Association Educational Foundation, used under license by National Restaurant Association Solutions, LLC a wholly owned subsidiary of the National Restaurant Association.

Copyright © 2009 National Restaurant Association Educational Foundation. Published by Pearson Education, Inc., Upper Saddle River, New Jersey 07458. All rights reserved. Printed in the United States of America. No part of this document may be reproduced, stored in a retrieval system, or transmitted in any form or by any other means electronic, mechanical, photocopying, recording, scanning or otherwise, except as permitted under Section 107 and 108 of the 1976 United States Copyright Act, without prior written permission of the publisher.

Requests for permission to use or reproduce material from this book should be directed to:
Copyright Permissions
National Restaurant Association Solutions
175 West Jackson Boulevard, Suite 1500
Chicago, IL 60604
Phone: 312-715-1010   Fax: 312-566-9729
Email: permissions@restaurant.org

5 6 7 8 9 V036 12 11 10
ISBN-13: 978-0-13-501898-9
ISBN-10:   0-13-501898-6

Contents

How to Take the ManageFirst Examination	1
Chapter Summaries and Objectives	9
Menu Marketing and Management Practice Questions	17
Answer Key	27
Explanations to Answers	29
Glossary	41

How to Take the ManageFirst Examination

The ability to take tests effectively is a learned skill. There are specific things you can do to prepare yourself physically and mentally for an exam. This section helps you prepare and do your best on the ManageFirst Examination.

I. BEFORE THE EXAM

A. How to Study
Study the right material the right way. There is a lot of information and material in each course. How do you know what to study so you are prepared for the exam? This guide highlights what you need to know.

1. **Read the Introduction to each *Competency Guide*.** The beginning section of each guide explains the features and how it is organized.

2. **Look at how each chapter is organized and take clues from the book.**

   - *The text itself is important.* If the text is bold, large, or italicized you can be sure it is a key point that you should understand.

   - *The very first page tells you what you will learn.*

     Inside This Chapter: This tells you at a high level what will be covered in the chapter. Make sure you understand what each section covers. If you have studied the chapter but cannot explain what each section pertains to, you need to review that material.

Learning Objectives: After completing each chapter, you should be able to accomplish the specific goals and demonstrate what you have learned after reading the material. The practice exam as well as the actual exam questions relate to these learning objectives.

- *Quizzes and Tests*

    Test Your Knowledge: This is a pretest found at the beginning of each chapter to see how much you already know. Take this quiz to help you determine which areas you need to study and focus on.

- *Key Terms* are listed at the beginning of each chapter and set in bold the first time they are referred to in the chapter. These terms—new and specific to the topic or ones you are already familiar with—are key to understanding the chapter's content. When reviewing the material, look for the key terms you don't know or understand and review the corresponding paragraph.

- *Exhibits* visually depict key concepts and include charts, tables, photographs, and illustrations. As you review each chapter, find out how well you can explain the concepts illustrated in the exhibits.

- *Additional Exercises*

    Think About It sidebars are designed to provoke further thought and/or discussion and require understanding of the topics.

    Activity boxes are designed to check your understanding of the material and help you apply your knowledge. The activities relate to a learning objective.

- *Summary* reviews all the important concepts in the chapter and helps you retain and master the material.

3. **Attend Review Sessions or Study Groups**. Review sessions, if offered, cover material that will most likely be on the test. If separate review sessions are not offered, make sure you attend class the day before the exam. Usually, the instructor will review the material during this class. If you are a social learner, study with other students; discussing the topics with other students may help your comprehension and retention.

4. **Review the Practice Questions,** which are designed to help you prepare for the exam. Sample questions are designed to familiarize the student with the format, length, and style of the exam questions, and represent only a sampling of topic coverage on the final exam. The performance level on sample questions does not guarantee passing of a ManageFirst Program exam.

B. *How to Prepare Physically and Mentally*

Make sure you are ready to perform your best during the exam. Many students do everything wrong when preparing for an exam. They stay up all night, drink coffee to stay awake, or take sleep aids which leave them groggy and tired on test day.

There are practical things to do to be at your best. If you were an athlete preparing for a major event, what would you do to prepare yourself? You wouldn't want to compete after staying up all night or drinking lots of caffeine. The same holds true when competing with your brain!

1. **Get plenty of sleep.** Lack of sleep makes it difficult to focus and recall information. Some tips to help you get a good night's sleep are:

   - Make sure you have studied adequately enough days before the exam so that you do not need to cram and stay up late the night before the test.
   - Eat a good dinner the night before and a good breakfast the day of the exam.
   - Do not drink alcohol or highly-caffeinated drinks.
   - Exercise during the day, but not within four hours of bedtime.
   - Avoid taking sleep aids.

2. **Identify and control anxiety.** It is important to know the difference between actual test anxiety and anxiety caused by not being prepared.

   Test anxiety is an actual physical reaction. If you know the information when you are **not** under pressure but feel physically sick and cannot recall information during the exam, you probably suffer from test anxiety. In this case, you may need to learn relaxation techniques or get some counseling. The key is how you react under pressure.

   If you cannot recall information during reviews or the practice exam when you are not under pressure, you have not committed the information to memory and need to study more.

   - Make sure you are as prepared as possible. (See "Anxiety Caused by Lack of Preparation")
   - Take the exam with a positive attitude.
   - Do not talk to other students who may be pessimistic or negative about the exam.
   - Know what helps you relax and do it (chewing gum, doodling, breathing exercises).
   - Make sure you understand the directions. Ask the instructor questions *before* the test begins.
   - The instructor or proctor may only talk to you if you have defective materials or need to go to the restroom. They cannot discuss any questions.
   - The instructor or proctor may continuously monitor the students so do not be nervous if they walk around the room.
   - Know the skills described in Section II, During the Test.

3. **Anxiety Caused by Lack of Preparation.** The best way to control anxiety due to lack of preparation is focus on the exam. Whenever possible, you should know and do the following:

   - Know the location of the exam and how to get there.
   - Know if it is a paper-and-pencil test or an online exam. Pencils may be available but bring sufficient number 2 pencils if taking the paper-and-pencil version of the exam.
   - If it is an online exam you may need your email address, if you have one, to receive results.
   - You are prohibited from using purses, books, papers, pagers, cell phones, or other recording devices during the exam.
   - Calculators and scratch paper may be used, if needed. Be sure your calculator is working properly and has fresh batteries.
   - The exam is not a timed; however, it is usually completed in less than two hours.
   - Take the sample exam so you know what format, style, and content to expect.
   - Arrive early so you don't use valuable testing time to unpack.

II. DURING THE TEST

An intent of National Restaurant Association Solutions' ManageFirst exams is to make sure you have met certain learning objectives. If you are physically prepared, have studied the material, and taken the practice exam, you should find the ManageFirst exams to be very valid and fair. Remember, successful test taking is a skill. Understanding the different aspects of test preparation and exam taking will help ensure your best performance.

A. *Test Taking Strategies*

- Preview the exam for a quick overview of the length and questions.
- Do not leave any question unanswered.
- Answer the questions you are sure of first.

- Stop and check occasionally to make sure you are putting your answer in the correct place on the answer sheet. If you are taking an online exam, you will view one question at a time.
- Do not spend too much time on any one question. If you do not know the answer after reasonable consideration, move on and come back to it later.
- Make note of answers about which you are unsure so you can return to them.
- Review the exam at the end to check your answers and make sure all questions are answered.

B. Strategies for Answering Multiple-Choice Questions

Multiple-choice tests are objective. The correct answer is there, you just need to identify it.

- Try to answer the question before you look at the options.
- Use the process of elimination. Eliminate the answers you know are incorrect.
- Your first response is usually correct.

III. AFTER THE EXAM

Learn from each exam experience so you can do better on the next. If you did not perform on the exam as you expected, determine the reason. Was it due to lack of studying or preparation? Were you unable to control your test anxiety? Were you not focused enough because you were too tired? Identifying the reason allows you to spend more time on that aspect before your next exam. Use the information to improve on your next exam.

If you do not know the reason, you should schedule a meeting with the instructor. As all NRA Solutions ManageFirst exams are consistent, it is important to understand and improve your exam performance. If you cannot identify your problem areas, your errors will most likely be repeated on consecutive exams.

IV. EXAM DAY DETAILS

The information contained in this section will help ensure that you are able to take the exam on the scheduled test day and that you know what to expect and are comfortable about taking the exam.

- Have your photo identification available.
- Anyone with special needs must turn in an *Accommodation Request* to the instructor at least 10 days prior to the exam to receive approval and allow time for preparations. *If needs are not known 10 days prior, you may not be able to take the exam on the scheduled test day.*
- A bilingual English-native language dictionary may be used by anyone who speaks English as a second language. The dictionary will be inspected to make sure there are no notes or extra papers in it.
- If you are ill and must leave the room after the exam has begun you must turn in your materials to the instructor or proctor. If you are able to return, your materials will be returned to you and you may complete the exam. If it is an online exam you must close your browser and if the exam has not been graded yet, login in again when you return.
- Restroom breaks are allowed. Only one person may go at a time and all materials must be turned in prior to leaving the room and picked up when you return; or you must close your browser and login again for online exams.
- Make-up tests may be available if you are unable to take the exam on test day. Check with your instructor for details.
- If you are caught cheating you will not receive a score and must leave the exam location.

Menu Marketing and Management Chapter Summaries and Objectives

Chapter 1 Factors that Impact Menu Item Selection

Summary

Planning a menu involves identifying and selecting menu items for inclusion in a restaurant's menu. Aspects to consider include the target market, the sales potential of the proposed menu items, other restaurants that will compete for the same customers, current restaurant industry trends, and consumer trends. When developing a restaurant menu, it is important to understand the marketing environment in which it exists, including an assessment of competitors' menu offerings and pricing parameters. Consumer trends include a demand for healthier menu offerings, Asian fare and quick, family meals offered at a reasonable price. In addition, the restaurant industry is very competitive, and one of the tools to success is to create a strong brand image in the minds of target customers. A well-planned menu and its superb operational execution help to reinforce the restaurant's brand image, leading to financial success.

The decision to include or exclude a menu item may depend upon more than customer tastes and preferences. It may also depend upon whether or not the item can be priced to meet the targeted profit margin and price point category the restaurant uses. Successfully executing a menu also must take into account operating efficiency and staff capability.

After completing this chapter, you should be able to:
- Define foodservice terms related to menus.
- Identify elements of the marketing environment.
- List factors that impact menu item selection.
- Select menu items.
- Discuss the impact of internal concerns when selecting menu items.

Chapter 2 Meeting the Nutritional Needs and Food Preferences of Customers

Summary

It is very important to recognize that customers select menu items for a variety of reasons, including hunger, nutrition, indulgence, dietary restriction, price, value, convenience, and marketing. Customers also have expectations of menu choices that will help them meet their nutritional needs for carbohydrates, fats, and proteins. Receiving, storing, handling, preparing, cooking, and serving food items appropriately will help maintain their nutritional quality.

Customers may have needs for special types of food because they are vegetarians, have allergies to certain food components, or have other dietary restrictions. The FDA has issued requirements for providing nutritional information to consumers on packaged food items and for any dietary or health claims made on a menu.

After completing this chapter, you should be able to:
- Outline the factors that influence food item selection by customers.
- Identify the sources of carbohydrates, proteins, and fats on the menu.
- Provide nutritional information to customers.
- Identify preparation and cooking methods that preserve nutrients in quantity cooking.
- Recognize the various types of vegetarian diets.
- Identify procedures for preparing food items for customers with allergies.

Chapter 3 Menu Layout and Design

Summary

The menu is the most important sales and marketing tool used in a restaurant. There are design and layout principles that help make the menu more effective as a sales tool. A good menu provides accurate information to the guest, is easy to read, is manageable in size and shape, and reinforces the brand identity of the restaurant. The menu is used for many purposes, some internal and some external. The menu provides a list of products and services offered to the customer and is an effective point-or-purchase merchandising tool. Internally, it is used for purchasing, production, and service. A good menu reinforces your relationship with your target market and serves as an advertising tool that your customer will read. It can be used externally if the menu is distributed to the public. A menu can be designed to influence customer purchase of the items you want to sell through the use of layout and pricing psychology.

After completing this chapter, you should be able to:
- List and describe the purposes of a menu.
- Explain how the menu communicates and reinforces the foodservice operation's brand.
- Explain how the menu reinforces marketing.
- Explain how the menu is a powerful sales tool.
- Explain how the menu can guide customers to select desired items.
- Explain how to use menu layout and pricing psychology to influence customer purchases.
- Explain the principles of menu layout and design.

Chapter 4 Menu Pricing

Summary

Price is an important part of the strategic marketing plan for a restaurant. It serves the purpose of determining profitability and providing information to the customer. Price has an important role in brand positioning, setting pricing objectives, and achieving the intended gross profit margin. The external environment, from the local competition to global events, is important to consider when pricing menus.

Price can be used to differentiate a competitive strategy from a premium pricing strategy. Consumers are sensitive to price and expect value for the money spent in the restaurant. The market can be segmented into specific market segments or target markets using pricing strategies.

Four basic pricing methods are used to determine menu prices: the food cost percentage, the average check method, the contribution margin method, and the straight markup pricing method using a dollar amount or a percentage.

Ways to account for employee meals include providing different food for employees, assigning a fixed dollar amount for each employee meal, and accounting for employee meal food cost using the restaurant in-house ordering system.

After completing this chapter, you should be able to:
- Explain the meaning of price and its use in strategic marketing.
- Describe the impact of the external environment on price.
- Outline a variety of pricing strategies used by an organization.
- List, compare, and contrast basic pricing methods.
- Explain the relationship of gross profit margin and profitability to pricing menu items.
- Explain methods for costing employee meals.

Chapter 5 The Alcohol Beverage Menu

Summary

Offering alcoholic beverages to the customer is an important part of many restaurants' operations. It is also important to observe the state and local laws applying to the sale and service of alcoholic beverages. Ways to present the liquor menu to the guest include offering the beer, wine, and spirits on the food menu, or on separate menus if the lists are extensive. Dessert menus may include desserts, after-dinner drinks, such as brandies and liqueurs, and specialty coffees. Methods for listing wines and spirits include indicating the distinctive features of each, such as brand name or vintner, year of harvest or production, descriptions of the wine or drinks made from the spirits, and suggestions of food items that would be enhanced by a particular wine. Wines include table wines, fortified wines, sparkling wines, aperitifs, and dessert wines. Distilled spirits include grain spirits, plant liqueurs and cordials are interchangeable terms. Beer includes lagers and ales. Lagers make up the majority of familiar beer in the United States. Ways to price alcoholic beverages include gross profit targets on larger quantities like bottles and markup pricing for glasses and mixed drinks.

After completing this chapter, you should be able to:
- Identify and list ways to present a liquor menu.
- List and describe typical elements of wine and spirits lists and the methods used to merchandise them.
- Categorize and describe spirits, beers, ales, liqueurs, and cordials.
- Price the beverage menu.

Chapter 6 Menu Item Sales Performance Analysis

Summary

A menu planner spends a lot of time developing a restaurant menu. Market research is done, the brand image is created, the menu items are selected and priced, the kitchen and storage areas are examined, and the menu layout is designed. This chapter has provided some tools to use in analyzing the performance of individual menu items in terms of customer satisfaction and profitability. The sales volume, or popularity index, indicates which items are the most popular and sell the most. The item contribution margin, or gross profit, indicates each item's individual profitability. Total contribution margin takes both of these factors into account to give a broader picture of each item's contribution to the foodservice operation's success. The above calculated figures can be calculated for the whole day, but are more accurate when done for separate day parts. Other considerations include item counts and the effect of stressing the 20 percent of items that generate 80 percent of revenue. But only taking into account these financial figures for menu items ignores other critical factors such as meeting the needs of customers, reinforcing the brand image, and maintaining kitchen efficiency. All these factors must be considered to make sound decisions about menu changes.

After completing this chapter, you should be able to:
- Analyze menu item sales performance.
- Calculate sales volume percentage and sales dollar percentage.
- Define profitability and target contribution margins.
- Analyze and evaluate the menu using item counts, subjective evaluation, popularity indexes, contribution margin, and day-part information.

Chapter 7 Menu Sales Mix Analysis

Summary

A sales mix analyzes the popularity and the profitability of a group of menu items. One of the major purposes for performing a menu sales mix analysis is to monitor the effectiveness of the menu items to maximize profits. Another purpose is to determine which items have the highest sales.

There are seven steps to performing a sales mix analysis. The first step is to select items on the menu to be compared in the analysis. The most accurate analysis will focus on one menu category, not on the entire menu. The next step is to determine the menu mix percentage (MM%), also known as the popularity index or sales volume percentage, of each item by dividing the number of specific items sold by the total number of items sold.

The third step is to determine the menu item contribution margin or gross profit, which is the difference between the item's good cost and its menu selling price. After finding the individual item contribution margin, the total item contribution for each menu item can be found by multiplying the number of menu items sold by the contribution margin for the item an alternative method for calculating the total menu contribution margin is to determine the total revenue of all menu items and the total food costs of all the menu items, and calculating the differences between them.

Once the total contribution margin for each item has been determined, each menu item can be compared to the average contribution margin, or the total contribution margin divided by the total number of items sold. Whether or not the item's contribution margin falls above or below the average determines if it considered high or low.

The next step in the menu sales mix analysis is to classify the menu items as either stars, plow horses, puzzles, or dogs. Classifications are made by looking at the menu mix percentage and the contribution margin for each item. Decisions concerning a menu item's profitability can be made once items are classified into these categories.

There are alternative ways of doing a sales mix analysis, including the Miller matrix, with is more useful in quick-service and low average check operations, and the cost-margin analysis, which combines the food costs from the Miller matrix and the contribution margin concept from menu engineering.

After completing this chapter, you should be able to:
- Perform a menu sales mix analysis.
- Describe other purposes for a menu sales mix analysis.
- Determine menu item's popularity.
- Determine menu item's profitability.
- Classify menu items as stars, plow horses, puzzles, or dogs.
- Change the menu based on the results of the menu sales mix analysis.

Menu Marketing and Management Practice Questions

Please note the numbers in parentheses following each question. They represent the chapter and page number, respectively, where the content in found in the ManageFirst Competency Guide.

IMPORTANT: These sample questions are designed to familiarize the student with format, length and style of the examination questions, and represent only a sampling of topic coverage.

The grid below represents how the *actual* exam questions will be divided across content areas on the corresponding ManageFirst Program exam.

Menu Marketing and Management	1.	Factors That Impact Menu Item Selection	8
	2.	Meeting the Nutritional Needs and Food Preferences of Customers	9
	3.	Menu Layout and Design	14
	4.	Menu Pricing	14
	5.	The Alcohol Beverage Menu	8
	6.	Menu Item Sales Performance Analysis	11
	7.	Menu Sales Mix Analysis	11
		Total No. of Questions	**75**

The performance level on sample questions does not guarantee passing of a ManageFirst Program examination. Further, the distribution of sample exam questions with their focus on particular areas of subject matter within a ManageFirst Competency Guide is not necessarily reflective of how the questions will be distributed across subject matter on the actual correlating ManageFirst exam.

1. What is the menu sales mix analysis method introduced by David Pavesic that categorizes menu items by popularity, food cost and contribution margin? (7, 140)
   A. Total menu revenue
   B. Cost-margin analysis
   C. Menu mix percentage
   D. Miller matrix method

2. What is the price-value relationship? (2, 20)
   A. The promptness of service of food items, ease of consumption, and cleanup
   B. The cost of goods sold compared to food cost
   C. The customer's estimate of whether or not a received product or service meets their expectations
   D. The pricing of items in order to make a certain targeted level of profit

3. When adding a new menu item, consideration should be given to what part of the production staff? (1, 13)
   A. Wages
   B. Hours
   C. Desires
   D. Skill level

4. What does the term operational execution describe? (1, 8)
   A. How a menu is laid out to influence a customer's decision making process
   B. How the servers operated to correctly prepare food for a large banquet
   C. How all parts of the restaurant work together to provide the customer with a positive dining experience
   D. How a market environment is changed by consumer trends

5. Using the information below, what is the menu mix percentage percentage for sirloin steak? (7, 128)

Menu Item	# Sold	Selling Price	Item Food $
Sirloin Steak	23	$17.99	$ 9.25
Roast Chicken	47	$14.99	$ 6.50
Pasta	30	$12.99	$ 4.25

A. 23%
B. 47%
C. 117%
D. 230%

6. According to the Competency Guide, what is a target market? (1, 3)
A. A group of individual restaurants that come together to get group purchasing discounts
B. Those things used to identify the goods and services offered by a restaurant to differentiate the establishment from its competitors
C. A group of people with similar characteristics and similar demands of the marketplace
D. The dollar amount customers are willing to pay for menu items during a specific meal period

7. According to the Competency Guide, what are the two general categories of beers? (5, 94)
A. Stouts and lagers
B. Ales and stouts
C. Porters and ales
D. Lagers and ales

8. What classification of alcohol has an alcohol content from 11 to 24%? (5, 92)
A. Spirits
B. Ales
C. Wines
D. Lagers

9. What is a porter? (5, 95)
    A. Beer drawn from a tap and propelled by carbonation from a keg.
    B. Ale that is characterized by a strong, bittersweet flavor.
    C. Traditional German beer that is richer and sweeter in flavor than most lagers.
    D. A heavier, dark brown, strongly flavored ale produced from malt roasted at high temperatures.

10. Which is a good source of carbohydrates? (2, 21)
    A. Meat
    B. Rice
    C. Nuts
    D. Eggs

11. According to the Competency Guide, what is the main disadvantage of offering an extensive menu? (1, 8)
    A. Controlling costs and quality
    B. Difficulty in attracting a large customer base
    C. Limited number of menu items
    D. Lack of variety of choices

12. According to the Competency Guide, what can offering a kid's menu reinforce for a restaurant? (1, 7)
    A. Profit margin
    B. Menu price
    C. Brand image
    D. Server's attitude

13. A restaurant that is open only for breakfast is concentrating on which part of market segmentation? (3, 41)
    A. Age
    B. Special interest
    C. Income
    D. Time of day

14. According to the Competency Guide, why might operating costs in one location versus another location may be higher? (4, 60)
    A. Federal legislative mandates
    B. Varying access to supplies
    C. National boycott of products
    D. Higher national gas prices

15. Using the information below, what is the total contribution margin for this group of menu items? (6, 114)

Menu Item	# Sold	Selling Price	Item Food $
Sirloin Steak	26	$17.90	$ 9.24
Beef Tips	23	$14.90	$ 6.48
Pork Roast	43	$15.90	$ 5.80
Lamb Chops	22	$20.90	$10.54
BBQ Ribs	39	$17.90	$ 9.41
Roast Chicken	51	$15.90	$ 7.54
Tuna Steak	21	$22.90	$ 9.18
Pasta	59	$16.90	$ 6.17

    A. $2,297.07
    B. $2,759.70
    C. $4,966.95
    D. $7,263.25

16. What are common food allergens found in restaurant kitchens on a daily basis? (2, 31)
    A. Fish and shellfish
    B. Beef and pork
    C. Cucumbers and carrots
    D. Turkey and chicken

17. What does the USDA recommend that a two thousand-calorie a day diet should contain? (2, 22)
    A. Twenty percent of saturated fats
    B. Two and one-half cups of vegetables
    C. One cup of fat-free milk or milk products
    D. Less than three servings of whole grains

18. According to the Competency Guide, where is the best place for the price of an item to appear on a menu? (3, 44)
    A. To the left of the item's name
    B. After the description of the item
    C. Above the description of the item
    D. In bold print before and after the item description

19. Where should the item a restaurant desires to be its most favored item be placed on a menu? (3, 43)
    A. Center of the page
    B. Bottom of the page
    C. Upper left of the page
    D. Upper right of the page

20. A vegan is a person who eats only what? (2, 30)
    A. Dairy products and eggs
    B. Organic fruits, vegetables, nuts, seeds, and eggs
    C. Grains, legumes, vegetables, fruits, nuts and seeds
    D. Vegetable and soy based foods and dairy products

21. What is the strictest type of vegetarian diet? (2, 30)
    A. Lacto-vegetarian diet
    B. Vegan diet
    C. Lacto-ovo-vegetarian diet
    D. Organic-vegetarian diet

22. How many 1.5 ounce drinks can be poured from a 25-ounce bottle of alcohol with a 5% loss to spillage? (5, 99)
    A. 18.4
    B. 15.8
    C. 15.3
    D. 16.1

23. What is the spirit cost of a drink with 2 ounces of alcohol from a bottle of scotch that cost $35.50 and has 22 salable ounces? (5, 99)
    A. $0.81
    B. $1.61
    C. $3.23
    D. $3.37

24. What is the common markup percentage of beer? (5, 100)
    A. 20-25 percent
    B. 40-45 percent
    C. 50-55 percent
    D. 70-75 percent

25. What is the menu price for a bottle of wine that costs $25.50 with a markup of 75%? (5, 97)
    A. $19.13
    B. $34.63
    C. $40.19
    D. $44.63

26. According to the Competency Guide, sales mix analysis is an analysis of the popularity and profitability of what? (7, 123)
    A. Competitors' menus
    B. Each menu item
    C. A group of menu items
    D. An entire menu

27. According to the Competency Guide, what is the element of marketing that focuses on the sale of goods and services to customers? (3, 40)
    A. Merchandising
    B. Competition
    C. Management
    D. Branding

28. What is product-bundle pricing? (4, 65)
    A. Combining several products and offering them at a price lower than the price would be if they were sold individually
    B. Pricing based on competitors' pricing structures because similar demands are being met
    C. Selling products at a higher price to create prestige or a premium position
    D. Refining markets into smaller groups with similar wants and needs

29. According to the Competency Guide, how can price cuts to increase sales volume be achieved? (4, 62)
    A. Reducing portion size
    B. Introducing kids' menus
    C. Substituting lower quality menu items
    D. Offering half-off coupons

30. What is the equation for the set dollar amount markup method? (4, 76)
    A. Food cost × Markup = Menu price
    B. Food cost + Markup = Menu price
    C. Food cost × Percentage = Markup
    D. Food cost + Contribution margin = Menu price

31. What is the equation for item contribution margin? (7, 130)
    A. Item selling price − Item food cost
    B. Number of item sold / Total number of all items sold
    C. Item's popularity index / Item's expected popularity
    D. Item sales dollars / Total sales dollars X 100

32. If an operation with a gross profit of $31,224, has direct labor costs of $10,204, controllable expenses of $8,653, and noncontrollable expenses of $7,772, what is its profit or loss? (4, 68)
    A. Profit of $4,595
    B. Profit of $3,897
    C. Loss of $2,894
    D. Profit of $12,367

33. On Thursday, a restaurant sold 198 entrées; 49 were BBQ Ribs and 26 were Roasted Chicken. What are the popularity indexes for BBQ Ribs and Roasted Chicken? (7, 128)
    A. BBQ Ribs 33%, Roasted Chicken 11%
    B. BBQ Ribs 25%, Roasted Chicken 13%
    C. BBQ Ribs 27%, Roasted Chicken 15%
    D. BBQ Ribs 42%, Roasted Chicken 12%

34. Using the information below, what is the popularity index for Chicken Vesuvio? (7, 128)

Entrée	Menu Price	Food Cost	# Sold
Chicken Vesuvio	$14.95	$ 6.29	22
Chicken Parmesan	$14.95	$ 6.90	17
Spaghetti Bolognese	$12.95	$ 5.12	33
Pork Tenderloin	$16.95	$ 7.15	31
Sirloin Steak	$18.95	$10.37	12
Salmon	$21.95	$13.72	11

A. 9%
B. 10%
C. 13%
D. 17%

35. According to the Competency Guide, with what number do mid-range dining operations' prices usually end? (3, 45)
A. 9
B. 0
C. 5
D. 2

Menu Marketing and Management Answer Key to Practice Questions

1.	B		26.	C
2.	C		27.	A
3.	D		28.	A
4.	C		29.	D
5.	A		30.	B
6.	C		31.	A
7.	D		32.	A
8.	A		33.	B
9.	D		34.	D
10.	B		35.	C
11.	A			
12.	C			
13.	D			
14.	B			
15.	B			
16.	A			
17.	B			
18.	B			
19.	A			
20.	C			
21.	B			
22.	B			
23.	C			
24.	A			
25.	D			

Menu Marketing and Management Explanations to the Answers for the Practice Questions

Question #1
Answer A is wrong. Total menu revenue is the number of each item sold multiplied by each menu item's selling price
Answer B is correct. Cost-margin analysis was introduced by David Pavesic and combines the food costs of the Miller matrix and the contribution margin concept from menu engineering.
Answer C is wrong. Menu mix percentage is found by dividing the number of specific items sold by the total number of items sold.
Answer D is wrong. Miller Matrix attempts to identify menu items that are low in food costs and popular.

Question #2
Answer A is wrong. The promptness of service of food items, ease of consumption, and cleanup is convenience.
Answer B is wrong. The cost of goods sold versus the item's food cost is the item contribution margin.
Answer C is correct. Customers do want value for the price they pay and the price-value relationship is a measure of whether or not their expectations of this value have been met.
Answer D is wrong. The pricing of items in order to make a certain targeted level of profit is profit-oriented pricing.

Question #3
Answer A is wrong. The wages of the production staff do not affect the ability to add a new item, so it does not need to be considered.
Answer B is wrong. The production staff's hours should not need to change in order to add one item to the menu.
Answer C is wrong. While it is important to keep the production staff happy, if the operation will benefit from adding a new item, the productions staff's desires do not need to be considered.
Answer D is correct. The skill level of the production staff is very important in deciding if a new item should be added or not, because they need to have the ability to produce all items on the menu.

Question #4
Answer A is wrong. The menu psychology deals with the way a menu is laid out to influence a customer's decision making process.
Answer B is wrong. Servers do not prepare food. That task is usually performed by back-of-house employees.
Answer C is correct. Operational execution is the term used to describe how all parts of a restaurant work together to provide the customer with a positive dining experience.
Answer D is wrong. Operations execution focuses on restaurant operations as opposed to consumer trends.

Question #5
Answer A is correct. Menu mix percentage of an item is the number of specific items sold divided by total number of all items sold. For sirloin steak this is 23 divided by 100 = .23 or 23%
Answer B is wrong. This answer, .47 or 47%, represents the menu mix percentage for roast chicken, not sirloin steak.
Answer C is wrong. If the menu mix percentage is the number of specific items sold divided by total number of all items sold, the menu mix percentage for sirloin steak is not 1.17 or 117%.
Answer D is wrong. If the menu mix percentage is the number of specific items sold divided by total number of all items sold, the menu mix percentage for sirloin steak is not 2.30 or 230%.

Question #6
Answer A is wrong. Group discounts apply to buying, not identifying customers.
Answer B is wrong. This answer describes the elements that make up a restaurant's brand image, not the target audience.
Answer C is correct. A target market is a group of people with similar characteristics and similar demands of the marketplace.
Answer D is wrong. This answer describes the price point, not the target audience.

Question #7
Answer A is wrong. Stouts is not a category of beers, but is a type.
Answer B is wrong. Stouts is not a category of beers, but is a type.
Answer C is wrong. Porters is not a category of beers, but is a type.
Answer D is correct. Lagers and ales are the major categories of beers.

Question #8
Answer A is wrong. Spirits have an alcohol content that varies from about 40 to 50%.
Answer B is wrong. Ale is a type of beer, with its alcohol content at about 3 to 7%.
Answer C is correct. Wine has an alcohol content of about 11 to 24%.
Answer D is wrong. Lager is a type of beer, with its alcohol content at about 3 to 7%.

Question #9
Answer A is wrong. Beer drawn from a tap and propelled by carbonation from a keg is draft beer.
Answer B is wrong. Ale that is characterized by a strong, bittersweet flavor is a stout.
Answer C is wrong. Bock beer is a traditional German beer that is richer and sweeter in flavor than most lagers.
Answer D is correct. Porter is heavier, dark brown, strongly flavored ale produced from malt roasted at high temperatures.

Question #10
Answer A is wrong. Meat is a good source of protein.
Answer B is correct. Grains such as breads, rice, cereal, and pasta are a good source of carbohydrates.
Answer C is wrong. Nuts are a good source of fat.
Answer D is wrong. Eggs are a good source of protein.

Question #11
Answer A is correct. With a larger menu, it is more challenging to control the costs and quality of choices.
Answer B is wrong. A variety of choices attracts a larger variety of customers.
Answer C is wrong. An extensive menu has many menu items.
Answer D is wrong. An extensive menu has a large variety of choices.

Question #12
Answer A is wrong. The profit margin may be affected by the availability of a Kid's menu, but is not reinforced by it.
Answer B is wrong. Menu prices vary from a regular menu and a Kid's menu, so the Kid's menu cannot reinforce regular menu prices.
Answer C is correct. The brand image is the type of operation a restaurant is trying to have, and by having a Kid's menu, the operation asserts itself as a family restaurant.
Answer D is wrong. A Kid's menu can be a nice menu offering but it is not indicative of a server's attitude in any way.

Question #13
Answer A is wrong. The time a restaurant is open is not an indication that it is concentrating on an age segment.
Answer B is wrong. Since breakfast in not a special interest, the restaurant is not concentrating on the special interest segment by being open only for breakfast.
Answer C is wrong. There is no indication as to what income a restaurant is targeting because it is open only for breakfast. Breakfast restaurants could range from value to upscale operations.
Answer D is correct. Since the restaurant focuses on a meal served in the morning, it is concentrating on the morning time of day segment.

Question #14
Answer A is wrong. Federal legislative mandates do not cause a difference in prices because they are applied across all geographical areas.
Answer B is correct. Access to supplies vary throughout the country and the harder they are to obtain in certain geographical areas, the more they will cost for that area.
Answer C is wrong. National boycotts affect all geographical areas.
Answer D is wrong. If the gas prices are national, then the products' transportation costs will be higher across all geographical areas.

Question #15
Answer A is wrong. The total contribution margin for the menu items does not equal $2,297.07.
Answer B is correct. The total contribution margin for the menu items equals $2,759.70. The total contribution margin for an item is determined by multiplying the contribution margin for each menu item by the number sold for that item.
Answer C is wrong. The total contribution margin for the menu items does not equal $4,966.95.
Answer D is wrong. The total contribution margin for the menu items does not equal $7,263.25.

Question #16
Answer A is correct. Fish and shellfish are common allergens.
Answer B is wrong. Beef and pork are not common allergens.
Answer C is wrong. Vegetables such as cucumbers and carrots are not common allergens.
Answer D is wrong. Eggs and eggs products are common allergens, but not turkey and chicken.

Question #17
Answer A is wrong. Less than ten percent of a diet should be saturated fats.
Answer B is correct. Two and one-half cups of vegetables are recommended by the USDA.
Answer C is wrong. Three cups of fat-free milk or milk products are recommended per day.
Answer D is wrong. A minimum of three servings of whole grains is recommended per day.

Question #18
Answer A is wrong. If the price were placed to the left of the item's name, that would be the first thing a customer sees and would affect their choice too much.
Answer B is correct. By placing the price after the description of the item, it helps to keep the price from being the first thing the customer sees and from affecting his or her choice too much.
Answer C is wrong. The customer's choice should be affected more by the description of the item rather than the price, so the description of the item should be first.
Answer D is wrong. Bold print would draw too much attention to the price and would prevent the item's description from most affecting the customer's choice.

Question #19
Answer A is correct. On a single-page menu, a customer's eye initially goes up from the center and then near the bottom of the page, so the first item you want the customer to see should be placed at the center, since this is the favored position.
Answer B is wrong. The customer does not look at the bottom of the page until after the customer looks at the center and at the top.
Answer C is wrong. The customer does not look at the top of the page until after the customer looks at the center.
Answer D is wrong. The customer does not look at the top of the page until after the customer looks at the center.

Question #20
Answer A is wrong. Vegans do not eat dairy or eggs.
Answer B is wrong. Vegans do not necessarily eat only organic foods and they do not eat eggs.
Answer C is correct. Vegans will not eat anything with animal products or by-products.
Answer D is wrong. Vegans do not eat dairy products.

Question #21
Answer A is wrong. Lacto-vegetarians eat dairy products, which is an animal by-product.
Answer B is correct. Vegans eat no animal by-products at all, including eggs and dairy.
Answer C is wrong. Lacto-ovo-vegetarians eat dairy products and eggs, which are animal by-products.
Answer D is wrong. Organic-vegetarians eat only organic foods, but may or may not eat animal by-products, such as dairy products or eggs.

Question #22
Answer A is wrong. $25 - (25 \times .05) / 1.5 \neq 18.4$
Answer B is correct. First find the set loss per bottle. Subtract that figure from the total amount in the bottle. That amount divided by serving size yields the number of pours per bottle.
$\quad\quad 25 - (25 \times .05) / 1.5 = 15.8$
Answer C is wrong. $25 - (25 \times .05) / 1.5 \neq 15.3$
Answer D is wrong. $25 - (25 \times .05) / 1.5 \neq 16.1$

Question #23
Answer A is wrong. $\$35.50 / (22/2) \neq \$0.81$
Answer B is wrong. $\$35.50 / (22/2) \neq \$1.61$
Answer C is correct. Spirit cost per drink is equal to the cost of the bottle divided by the pours per bottle. $\$35.50 / (22/2) = \$3.23$
Answer D is wrong. $\$35.50 / (22/2) \neq \$3.37$

Question #24
Answer A is correct. The common markup percentage of beer is 20–25%.
Answer B is wrong. The common markup percentage of beer is 20–25%, not 40–45%.
Answer C is wrong. The common markup percentage of beer is 20–25%, not 50–55%.
Answer D is wrong. The common markup percentage of beer is 20–25% not 70–75%.

Question #25
Answer A is wrong. The formula to calculate the price of the bottle of wine is wine bottle cost X 100% + markup %. $25.50 + ($25.50 × .75) ≠ $19.13

Answer B is wrong. The formula to calculate the price of the bottle of wine is wine bottle cost X 100% + markup %. $25.50 + ($25.50 × .75) ≠ $34.63

Answer C is wrong. The formula to calculate the price of the bottle of wine is wine bottle cost X 100% + markup %. $25.50 + ($25.50 × .75) ≠ $40.19

Answer D is correct. The formula to calculate the price of the bottle of wine is wine bottle cost X 100% + markup %. Therefore, $25.50 + ($25.50 × .75) = $44.63

Question #26
Answer A is wrong. A sales mix analysis is used to analyze a single operation, not an operation versus its competition.

Answer B is wrong. A sales mix analysis analyzes a group of menu items, not just one item.

Answer C is correct. A sales mix analysis analyzes a group of menu items.

Answer D is wrong. A sales mix analysis analyzes a group of menu items, not an entire menu.

Question #27
Answer A is correct. Merchandising is concerned with the sale of goods and services to customers and includes advertising, product display, pricing, discounts, special offers, the invention of sales pitches, and identification of avenues for sales.

Answer B is wrong. Competition should be considered in how to approach merchandising, but is not an actual element of marketing.

Answer C is wrong. Management is concerned with the organization of goods and services, but not with their sales.

Answer D is wrong. Branding is important in determining the types of goods and services offered, but it is not concerned with the sale of these goods and service.

Question #28
Answer A is correct. Product-bundle pricing is combining products and selling them together for less than they would be sold for individually.
Answer B is wrong. Pricing based on competitors' pricing structures because similar demands are being met is competitive pricing.
Answer C is wrong. Selling products at a higher price to create prestige or a premium position is premium pricing.
Answer D is wrong. Refining markets into smaller groups with similar wants and needs is market segmentation.

Question #29
Answer A is wrong. A price cut is a decrease in price, not in the portion size because a decrease in portion size does not increase sales volume.
Answer B is wrong. Kids' menus offer smaller portions at a lesser cost, but are not price cuts because they do not necessarily increase sales volume.
Answer C is wrong. While price costs can cause perceived decrease in quality, a decrease in quality is not a form of price cutting.
Answer D is correct. Coupons can be offered as a price cut for customers.

Question #30
Answer A is wrong. The equation for set dollar amount markup is Food cost + Markup = Menu price, not Food cost × Markup = Menu Price
Answer B is correct. The equation for set dollar amount markup is Food cost + Markup = Menu price.
Answer C is wrong. The equation, Food cost × Percentage = Markup, is the set percentage increase method.
Answer D is wrong. The equation, Food cost + Contribution margin = Menu price, determines the menu price of an item.

Question #31
Answer A is correct. Item selling price − Item food cost = item contribution margin
Answer B is wrong. Number of item sold / Total number of all items sold = Menu mix percentage
Answer C is wrong. Item's popularity index / Item's expected popularity = Popularity factor
Answer D is wrong. (Item sales dollars / Total sales dollars) X 100 = Sales dollar percentage

Question #32
Answer A is correct. In order to determine a profit or loss, all expenses must be subtracted from the gross profit, and in this equation there is a profit of $4,595.
Answer B is wrong. In order to determine a profit or loss, all expenses must be subtracted from the gross profit, and in this equation there is not a profit of $ 3,897.
Answer C is wrong. In order to determine a profit or loss, all expenses must be subtracted from the gross profit, and in this equation there is not a loss of $2,894.
Answer D is wrong. In order to determine a profit or loss, all expenses must be subtracted from the gross profit, and in this equation there is not a profit of $12,367.

Question #33
Answer A is wrong. The popularity index is (Number of items sold / Total number of all items sold) × 100 and in this example (49 / 198) × 100 ≠ 33% and (26 / 198) × 100 ≠ 11%.
Answer B is correct. The popularity index is (Number of items sold / Total number of all items sold) × 100 and in this example (49 / 198) × 100 = 25% and (26 / 198) × 100 = 13%.
Answer C is wrong. The popularity index is (Number of items sold / Total number of all items sold) × 100 and in this example (49 / 198) × 100 ≠ 27% and (26 / 198) × 100 ≠ 15%.
Answer D is wrong. The popularity index is (Number of items sold / Total number of all items sold) × 100 and in this example (49 / 198) × 100 ≠ 42% and (26 / 198) × 100 ≠ 12%.

Question #34
Answer A is wrong. 9% is the popularity index for Salmon.
Answer B is wrong. 10% is the popularity index for Sirloin Steak.
Answer C is wrong. 13% is the popularity index for Chicken Parmesan.
Answer D is correct. The popularity index is (Number of items sold / Total number of all items sold) × 100 and for Chicken Vesuvio the equation is (22 / 126) × 100 = 17%.

Question #35
Answer A is wrong. Value pricing, quick-service operations' prices usually end in "9."
Answer B is wrong. Upscale dining operations' prices usually end in "0."
Answer C is correct. Mid-range value, casual dining operations' prices usually end in "5."
Answer D is wrong. Prices usually do not end in "2."

Menu Marketing and Management Glossary

80-20 rule   as used in this book, it is the suggestion that two or three appetizers (or other menu items in another category) often account for 80% or more of appetizer sales. More generally, the term suggests that a small number of causes create a large number of effects ("The vital few and the trivial many").

Ale   A beer made from top-fermenting yeast at warmer temperatures

Allergen   substance that produces an allergic reaction

Average contribution margin   determined by dividing the total item contribution margin by the total number of menu items sold

Bitters   Herbs or roots used as ingredients in other drinks

Bock Beer   A traditional German-style beer

Brand   those things used to identify the goods and services offered by a restaurant and to differentiate the establishment from its competitors

Brandy   A distilled spirit made from fruit

Call Brand   A specific brand of liquor

Call Liquor   Same as **call brand**

Calorie   amount of energy that a food contains

Carbohydrate   food group that consists primarily of sugars and starches

Cholesterol   lipid found in animal products that is considered a risk factor for heart attack and strokes in some people

Complete protein   proteins that contain the nine essential amino acids required by the human body

Contribution margin(1)   difference between a menu item's selling price and its food cost

Contribution margin(2)   combination of nonfood cost and profit

Contribution margin(3)   Food (or beverage) revenue – Food (or beverage) cost

Cordial   An alcoholic beverage made from grain, plant or fruit spirits and flavored with herbs, spices, fruits, nuts or other ingredients

Cost-margin analysis   sales mix analysis method that combines the food cost from the Miller Matrix method and the contribution margin concept from Menu Engineering

Cost of food sold   cost of goods sold; food cost

Day-part   period of a day that a particular meal or menu is served

Day-part analysis   process to determine which and how many items are sold at certain times of the day

Disposable income   amount of a person's income that can be used for discretionary spending after the household bills have been paid

Distilled Spirits   Alcoholic beverages made from fermented grains or other plants

Direct competitors   other restaurants in the same market targeting the same customers

Dog   name given to a menu item that is unpopular and unprofitable

Draft Beer   A beer drawn from a tap and propelled from a keg by carbonation

Expected popularity   popularity that a specific item in a group of menu items is anticipated to have by some authority

Fat   food group that provides approximately nine calories per gram and that is the most concentrated source of heat and energy

Food cost   same as **cost of food sold**

Food cost percentage   result of dividing the total food cost (cost of goods sold) by total sales

Generic Wine   Wines made from a blend of wines

Grain Spirits   Liquors including gin, vodka and whiskey

Gross profit   same as **contribution margin**

Gross profit margin   money remaining after the cost of goods sold has been subtracted from total sales; same as contribution margin

Hops   Clusters of blossoms from the hop vine used in making beer

Hydrogenation   process by which unsaturated fat becomes a saturated fat

Income statement   accounting report of the revenues taken in and the expenses incurred by an organization during a certain time period

Incomplete protein   proteins that do not include all of the essential amino acids

Indirect competitors   grocery stores and other businesses serving food and complete meals that are ready to eat

Item contribution margin   the difference between a menu item's food cost and its selling price

Item contribution margin category   the "high" or "low" profitability of a menu item determined by comparing the item's contribution margin to the menu's average contribution margin

Item count number of items in a list

Lacto-ovo-vegetarian vegetarian who consumes all items that a vegan will eat plus dairy products and eggs

Lacto-vegetarian vegetarian who will consume all items that a vegan will eat plus dairy products

Lager A beer made from bottom-fermenting yeast at cooler temperatures

Leading vertical spacing between lines of type on a menu

Lipid type of molecule related to the fat family; for example, cholesterol

Liqueurs Spirits made from grain, plants or fruits and flavored with herbs, spices, fruits, nuts or other ingredients

Malt An ingredient of beer made from the grains of barley

Malt Liquor A beer with a higher alcohol content than other beers

Margin(1) difference between a menu item's food cost and its selling price

Margin(2) same as **contribution margin**

Market segmentation refining a restaurant's markets into smaller groups with similar wants and needs

Marketing strategy method selected by an organization to meet the needs of the defined group of customers through the products or services provided, the means of communication, the distribution channels, and the price

Markup factor amount over a base (such as food cost) that a menu item is increased to establish a menu price

Mathematical price selling price of a menu item based only upon arithmetic calculations

Meat substitutes   items including vegetable- and soy-based burgers that many vegetarians will eat

Menu item classification   name given to a menu item based upon its profitability ("high" or "low") and popularity ("high" or "low")

Menu mix percentage   total sales of one menu item expressed as a percentage of the sales of all menu items

Menu mix percentage popularity rate   expected percentage of total sales that should be generated by the sale of one menu item The rate is calculated as follows: 100 ÷ Number of competing menu items × 0.70.

Menu selling price   selling price of a menu item as listed on the menu

Merchandising   element of marketing concerned with the sale of goods and services to customers

Miller matrix   menu sales mix analysis method that attempts to identify menu items that are low in food cost and that are popular

Minerals   these necessary nutrients in the human diet may be required in relatively large amounts (calcium, sodium, and potassium) or in trace amounts (iron, zinc, and iodine)

Must   Juice of grapes
Net loss   financial situation that arises when expenses are greater than revenues

Net profit   financial condition that arises when revenues exceed expenses

Nutrition   science of food, nutrients, and their actions in the body

Operational execution   term used to describe how all parts of the restaurant work together to provide the customer with a positive dining experience

Organic food food items grown by farmers that emphasize the use of renewable resources and the conservation of soil and water to enhance environmental quality for future generations

Pareto principle same as **80–20 rule**

Perishability likelihood of spoilage or decay

Plant Liquors Liquors including rum and tequila made from plants

Plow horse name given to a menu item that is popular but not profitable

Polyunsaturated type of unsaturated fat

Points units that measure the size of a letter or character; there are 72 points per inch.

Popularity factor calculation that makes comparisons between competing categories of menu items less difficult; Item's popularity index ÷ Item's expected popularity = Popularity factor

Popularity index(1) way to measure the popularity of a specific menu item in relation to other menu items in its category; (Item sales volume ÷ Sales volume of all category items) x 100 = Popularity index

Popularity index(2) same as **menu mix percentage**

Popular seller menu item that sells at a rate above the menu mix percentage popularity rate

Porter A heavier, dark brown, strongly flavored ale

Price point dollar amount customers are willing to pay for menu items during a specific meal period

Pricing objective   process used to obtain a selling price

Pricing strategy   same as **pricing objective**

Price-value relationship   customer's estimate of whether a product or service meets expectations

Product-bundle pricing   process of combining several products and offering them at a price lower than the price would be if they were sold individually

Profitability   excess of revenue over expenses in a series of transactions

Profit and loss statement   same as income and expense statement

Profit-oriented pricing   pricing strategies established to make a certain targeted level of profit

Proof   The number that represents twice the percentage of alcohol content of a distilled liquor

Protein   major food group required by the human body to promote growth, repair body tissue, and to help regulate body functions

Puzzle   name given to a menu item that is profitable but not popular

Recommended dietary allowance (RDA)   suggestions offered by the U.S. Department of Agriculture (USDA) about the types and quantities of different foods that should be consumed for good nutritional health

Sales dollar percentage   percentage, expressed in dollars, of sales that a menu item accounts for

Sales mix analysis   analysis of the popularity and the profitability of menu items in a group of menu items

Sales-oriented pricing   pricing strategies with the goal of maximizing sales volume not profit

Sales volume   number of times a menu item is sold during a specific time period

Sales volume percentage(1)   sales of a specific menu item expressed as a percentage of total sales of all menu items

Sales volume percentage(2)   same as **menu mix percentage**

Satiety value   satisfaction value

Saturated fat   type of fat generally found in animals (but cocoa butter and coconut oil also contain saturated fats)

Set dollar amount markup   straight markup pricing method that simply adds a fixed dollar amount to the food cost of an item

Set Loss   A percentage established by the restaurant for spirits to account for pouring mistakes, evaporation or spillage

Sommelier   A wine steward or person responsible for wine in a restaurant

Spirits   Another name for liquor or hard liquor

Standard Pour   The standard number of glasses of wine or (liquor) contained in a bottle

Star   name given to a menu item that is profitable and popular

Status quo pricing   pricing strategy designed to maintain one's competitive position relative to other restaurants in the market

Stout   A strong, bittersweet flavored ale

Straight markup   menu pricing method in which the selling price is obtained by marking up the cost according to a formula

Strategic marketing plan   description of markets and how a foodservice operation intends to approach them

Subjective evaluation   assessment based primarily on opinion

Target customers   group of people with similar characteristics and similar demands of the marketplace

Target margin   goal set by the management team about the restaurant's desired contribution margin

Target market   same as **target customers**

Target profit margin   net profit remaining to the investors and owners after all costs and business taxes have been paid

Total item contribution margin   total contribution margin generated from the sales of a single menu item

Total item food cost   total number of portions of one menu item sold multiplied by the food cost for one portion of the item

Total item revenue   total number of portions of a menu item sold multiplied by its selling price

Total menu contribution margin   total revenue generated from the sale of all menu items minus the total food costs incurred for those menu items

Total menu food cost   total food cost for all menu items sold

Total menu revenue   total of revenue generated from the sale of all menu items

Trademark   legal action that protects the brand name or symbol used by others

Trans fat   unsaturated fat that acts like a saturated fat in the body

Type size   size of a written letter or character

Type weight   thickness of a character

Typeface look and style of a letter

Unpopular seller menu items with a menu mix percentage below the menu mix percentage popularity rate

Unsaturated fat type of fat generally found in fish and vegetable oils

Value market low price segment of the market

Value pricing process of pricing products and services based upon their worth in usefulness or importance to the buyer

Varietal Wine Wines made from just one type of grape

Vegan vegetarian who follows the strictest diet of any vegetarian and consumes no dairy, eggs, meat, poultry, fish, or anything containing an animal product or by-product

Vegetarian person who consumes no meat, fish, or poultry products

Vintner Winery

Vitamins nutritional components that are a necessary part of the human diet. Water-soluble vitamins include vitamin C and B-complex vitamins. Fat-soluble vitamins include vitamins A, D, E, and K.

Wheat Beer Beer made from wheat grains

White space blank space

Wine Alcoholic beverage made by fermenting grapes

Yeast A living microorganism used in beer production to change sugars in grains and hops into carbon dioxide and alcohol